THE OPEN
MATHEMA[T]
AN INTERF[A]
MA290: TOP

BLOCK 1 MATHEMA[T]   [ANCIENT] WORLD

UNIT 2

MATHEMATICS IN THE GREEK WORLD

PREPARED BY JOHN FAUVEL FOR THE COURSE TEAM

THE OPEN UNIVERSITY

CONTENTS

2.1	A Greek Problem	3
2.2	Geometry Before Euclid: Chronology and Sources	5
2.3	Greek Mathematics: Classifications and Levels	12
2.4	Mathematical Traditions in the Hellenistic World	19
2.5	The Commentating Tradition	23

This unit forms part of an Open University course. The set book for the course, to which reference is made as **SB**, is:

John Fauvel and Jeremy Gray (editors), *History of Mathematics: A Reader*, Macmillan 1987.

Acknowledgements

Grateful acknowledgement is made to the following sources for material used in this unit: *Figure 4*, Staatliche Museen zu Berlin; *Figure 5*, Kunsthistorisches Museum, Vienna; *Figure 11*, Cambridge University Library; *Figure 10*, Tafel II, *Archiv für Papyrusforschung XVII* (1962) 1; *Figure 12*, Bodleain Library, D'Orville 301.

The Open University, Walton Hall, Milton Keynes.

First published 1987. Reprinted 1989, 1996, 1998, 2000, 2002, 2005.

Copyright © 1987 The Open University.

All rights reserved. No part of this publication may be reproduced, stored in a retrieval system or transmitted, in any form or by any means, without written permission from the publisher.

Designed by the Graphic Design Group of the Open University.

Typeset in Great Britain by Santype International Ltd, Salisbury.

Printed in the United Kingdom by Martins the Printers, Berwick upon Tweed.

ISBN 0 335 14246 X

This text forms part of the correspondence element of an Open University Second Level Course.

For general availability of supporting material referred to in this text, please write to Open University Educational Enterprises Limited, 12 Cofferidge Close, Stony Stratford, Milton Keynes, MK11 1BY, Great Britain.

Further information on Open University courses may be obtained from The Admissions Office, The Open University, P.O. Box 48, Milton Keynes, MK7 6AB.

1.6

2.1 A GREEK PROBLEM

In turning to Greek mathematics we move forward about a thousand years from Old Babylonian texts. By the sixth century BC, Greek-speaking peoples had settled around the eastern Mediterranean, from Sicily to Asia Minor, and from northern Africa to the Black Sea, living in relatively independent 'city-states' such as Athens, Sparta, Corinth and Syracuse. Despite political differences and varying forms of government, there were strong cultural bonds across the Greek-speaking world. This makes it possible for us to speak of 'Greek mathematics', meaning that the people with whom we are concerned spoke and wrote in the Greek language, but did not necessarily have much to do with the territory of modern Greece. Indeed, the earliest Greek scientific thought is associated with people who lived in Ionia—the western coast of what is now Turkey, across the Aegean Sea from Greece. But the best-known of Greek mathematical works, Euclid's *Elements*, was produced in Alexandria, in Egypt. In addition, although we are primarily concerned in this and the next two units with mathematical developments from about 450 to 200 BC, the Greek culture of which we are speaking lasted until around the sixth century AD, and was consciously imitated for several centuries thereafter. So our 'Greek mathematics' is widely spread both in space and time.

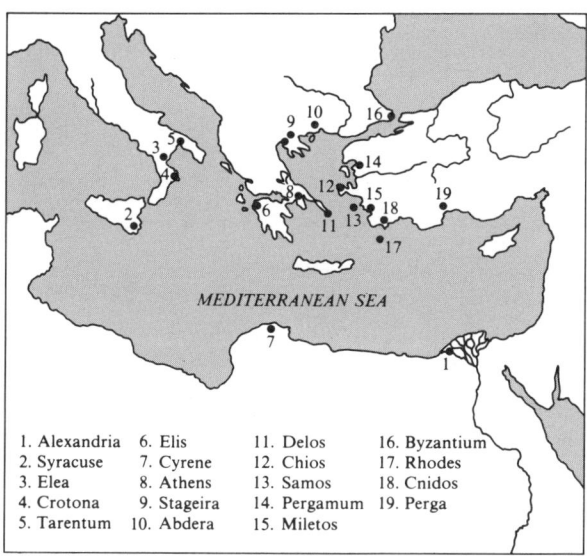

Figure 1 Some towns of the Greek-speaking world

We shall trace later the more detailed development of Greek mathematical thought, insofar as this can be determined from the sources. The first thing to notice is that Greek mathematical activity appears to differ in fundamental ways from Egyptian and Babylonian mathematics. To gain an initial perception of this, let us study the earliest available extended piece of reliable evidence about Greek mathematics, and work outwards from there. This evidence occurs in the course of Plato's dialogue *Meno*, dating from about 385 BC. The dialogue has two main characters, the philosopher Socrates (Plato's teacher) and Meno, a young aristocrat. They begin by discussing the nature of *virtue*, in particular, whether it is something that can be taught. In so doing, they move on to the more general question of whether knowledge of *anything* can be taught, and if so how. In trying to show that true knowledge can be elicited from someone who initially holds false beliefs, Socrates holds a mathematical discussion with a slave boy.

Question 1 Please *turn now to the passage 'Socrates and the slave boy'* in the Source Book (**SB** 2.E1), and read it in order to formulate answers to these three questions.

(i) What mathematical result have Socrates and the slave boy reached by the end of their conversation?

(ii) Sketch out a summary of the mathematical discussion, including the slave boy's mistaken beliefs.

(iii) What are the most notable ways in which the mathematical activity here appears to differ from what you have learnt of Babylonian and Egyptian mathematics?

Study Note This is your major activity for this section, so be prepared to spend up to an hour studying the passage and answering these questions, and then a further hour consolidating any points in the *Comment* which had not occurred to you. You might find it helpful to read through the passage fairly quickly first in order to answer (i), skipping anything you do not immediately follow. Then work through it more slowly in order to answer (ii), before sitting back and thinking about the impression you have gained from the whole passage and the way it contrasts with *Unit 1*.

Comment

(i) The problem was to find a square whose area was twice that of some initial square. The solution is to construct the square whose side is the diagonal of the first square.

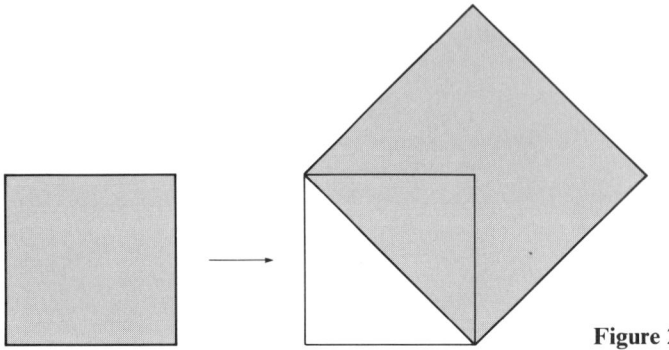

Figure 2

(ii) Having established that the given square, whose side is two feet, has an area of four feet, it follows that the square being sought must have an area of eight. The problem is, what side must it have? The slave boy says four: to double the square, double the side. But Socrates leads him to realise that this gives a square of area sixteen, which is not what was wanted. So the required side-length must lie between two and four. The slave boy tries three, but that gives an area of nine, still too big. The boy is somewhat perplexed by now, so Socrates draws another figure in the sand enabling him to see that the *diagonal* of the given square is the side wanted.

(iii) There are three main features here that you may have noticed. First, the character of the source: an articulate literary composition, in marked contrast to the staccato instructions and tables you have seen earlier. Of course, the contrast is partly attributable to the particular source, a philosophical dialogue written with more than purely mathematical intentions, but the impression remains nevertheless of a much fuller explanation of what is going on, and why, than you have encountered previously.

Secondly, the subject matter is different, being much more explicitly geometrical. That is, it is about the relationships of shapes, and such numbers as are introduced seem fairly incidental. Thus, notice how the slave boy's guesses at the numerical length of side required turn out to be the wrong approach. The solution comes about only once he has stopped guessing numbers and starts looking at a geometrical construction.

Thirdly, there is a different kind of concern in this passage, namely for showing *why* the result is true; care is taken throughout the argument to show what statements follow from other statements, and what are the implications of various conjectures. There is a logical air to the proceedings which we have not met before. The result comes about not so much by a series of instructions—'to double a square you do this, this and this'—but by a process whose point is to enable the boy to feel convinced that the construction leads to the right answer. Having seen the construction of the square on the diagonal, it becomes the boy's 'personal opinion' that this is the solution, because he has seen why, and Socrates goes on to say that when he has done this a few times, 'in the end he will have a knowledge on the subject as accurate as anybody's'. ∎

There are a number of further interesting points to emerge from this passage, which will help in our further exploration of Greek mathematics. We draw attention to them now, while the passage is fresh in your mind, continuing the numbering from the three major features just discussed.

(iv) Notice a curious feature towards the start of their conversation, namely that it takes the slave boy three answers to arrive at the result that the area of a square of side two is four. Now, the rest of the passage shows him to be quite quick and perceptive, albeit untutored mathematically, so there seems no reason to have laboured this point quite so much—*unless* the numerical determination of area was conceptually more subtle than we might initially have assumed.

(v) It is worth spelling out in more detail the *structure* of part of the mathematical argument. Let us take the boy's first belief, that to double the area you double the side. Socrates knows that this is wrong, but in order to *show* that it is wrong he goes along with the belief in order to pursue its implications. On the assumption of the side being doubled, he shows that this leads to 'not a double but a fourfold figure'. But that was not what was wanted, and hence the initial conjecture must have been false. Later we shall see more mathematical arguments with this structure.

(vi) Socrates makes a significant remark half-way through, after persuading the boy that his side of length three will not work either: 'Then what length will give it? Try to tell us exactly. If you don't want to count it up, just show us on the diagram.' The casual, conversational tone here is deceptive: Socrates is giving a strong hint that the length required *cannot* be told exactly, in numbers like two, three or four, but can be shown only on a diagram—the length can be pointed to, but not 'counted up'. So even the apparently straightforward procedure of attaching numbers to lengths of lines does not seem adequate for elementary geometrical investigations.

(vii) It would take us too far afield to explore all the philosophical overtones and setting of the passage, though you should note that this *is* its context. In particular, the distinction between having opinions and having knowledge is one to which we shall return later. Observe that in this full early source, mathematics comes in as exemplifying knowledge of an especially secure kind. We receive no hint of this conception in pre-Greek sources.

Now that our initial survey of this passage is over, you will see that quite a number of questions have been raised, puzzling features of which earlier, non-Greek sources gave no warning. Explicating these will take up much of the rest of this block. Of course, we cannot assume without further evidence that this source, apparently plucked out of thin air, is typical of the mathematical level or concerns even of its own period (early fourth century BC). But we can use it as a preliminary focusing of features in the Greek mathematical landscape that we should explore further.

2.2 GEOMETRY BEFORE EUCLID: CHRONOLOGY AND SOURCES

Where does our knowledge of Greek mathematics come from? Although Greek sources can be more chatty and informative than earlier ones, as you have seen, there is a problem in that what we have is very rarely the actual original. Whereas the Rhind Papyrus, in the British Museum, and the tablet Plimpton 322, in New York, are the physical objects that left the scribes' hands, there is no Greek equivalent of any significance. The Greeks generally wrote on papyrus like the Egyptians, and it does not weather well. What have come down to us are copies of copies of copies—so that it is not unusual for the earliest extant manuscript to be closer in time to us than to the original source. Furthermore in copying, people sometimes made mistakes, sometimes introduced further material or even left bits out. So establishing or reconstructing what the original text said is not always easy.

To learn more of how this is done, see **SB** 3.F1.

One implication of this is that the works which have come down to us are, broadly, only those which every successive generation has thought worthy of preservation, copying and handing on, for one reason or another. Conversely, there are many works whose names we know, and which we would now be very interested in studying, but which have not survived. For instance, Euclid's *Elements* (c. 300 BC) seems to have superseded several earlier works of the same name which are no longer extant: presumably they were discarded just as we discard outdated textbooks. Our evidence for the existence of these early *Elements* comes from a commentary on Euclid written by the scholar Proclus (c. AD 410–485). The historical survey of Greek mathematics that Proclus gives is the fullest we have that was written *relatively* near to the time; Proclus was born in Byzantium (now Istanbul) and educated in Alexandria and Athens, where he eventually became head of the Neoplatonic Academy. In reading his historical account of Greek mathematics until Euclid, it is worth bearing in mind that his time-relationship to what he describes is somewhat like ours to the Norman conquest or Robin Hood, so where Proclus found his information is an interesting question (but one that we shall not go into fully at present).

The word *Neoplatonic* refers to beliefs following the ideas of Plato. This is discussed in more detail later (Section 2.5).

Please *read now Proclus' summary* (**SB** 2.A1)—read it fairly swiftly for a general impression, and then a second time bearing Question 2 in mind, but ignore any unfamiliar names. (Of several of the names he cites, little more is known than told here by Proclus.)

Question 2

(i) Is it your impression that Proclus is attempting a fairly serious historical survey—that is, paying attention to what he can validly say in the light of the sources at his disposal? (In reaching your answer, you might like to study how Proclus reaches a decision about when Euclid lived, in the final paragraph.)

(ii) Proclus mentions, among others, the following people who will come into our story: Archimedes, Euclid, Eudoxus, Hippocrates, Plato, Pythagoras, Thales. For an initial understanding of the chronology, place these people in their time-ordering according to Proclus' account.

Comment

(i) Yes, there is a clear impression of Proclus using other sources and being careful to inform the reader of this. The start of his last paragraph mentions 'those who have written histories', which is generally taken to mean that he relied on these for the preceding account. His careful weighing of the evidence for when Euclid lived is both a tribute to his endeavour to be as truthful and accurate as he can, and also a reminder to us of how Euclid was already an inhabitant of the dim past when Proclus wrote.

(ii) The order given by Proclus (with which historians today would not disagree) is: Thales, Pythagoras, Hippocrates, Plato, Eudoxus, Euclid, Archimedes. ■

There are several interesting things arising from what Proclus says about these people and the development of geometry, which we shall return to later. For now, the important thing is to begin to form some fixed points on your chronological scale, into which your growing understanding of the development of Greek mathematics can be fitted as the block progresses. We take as the two 'fixed points' the two dates already acquired: the composition of Plato's *Meno* (385 BC) and that of Euclid's *Elements* (300 BC). (Both dates are approximate, best estimates in the light of all the evidence.) So our preliminary chronological division of Greek mathematics will be into three periods: before Plato (the sixth and fifth centuries BC); between Plato and Euclid (the fourth century BC); and after Euclid (from the third century BC onwards). This division is not, in fact, entirely arbitrary—Plato and Euclid were in their different ways important figures in the development of Greek mathematics, and there are good reasons, concerning what sources have survived, for framing our analysis this way. You can judge the validity of the approach better as we progress.

In the rest of this unit we shall survey the historical development of Greek mathematical activities, from our earliest records through to the transmission of classical Greek texts in the Byzantine Empire. Then, in the following two units, we shall return to fill out and analyse in more detail those aspects of the early period (up to the century after Euclid) during which the characteristic Greek approach to mathematics was formed and developed.

In the beginning, Proclus speaks of Greek geometry as being learnt from the Egyptians. Greek sources are surprisingly unanimous about this—surprisingly, because the Egyptian geometry that we know of bears little resemblance to Greek geometry (as exemplified, say, in the *Meno* passage or in Euclid's *Elements*), being much more in the nature of arithmetic applied to the measurement of shapes. So it is a little surprising that ancient authors saw a continuity of tradition between Egypt and Greece where it is the qualitative leap made by the Greeks that is more apparent to us.

You can read the earliest sources which describe this Egyptian origination of geometry in the Source Book (**SB** 1.D4)—notice that they agree on its Egyptian origin even while disagreeing on why it occurred.

In all events, the earliest Greek mathematical investigations that we hear of took place during the sixth century BC, associated with the name of Thales who lived at Miletos (see Figure 1), the wealthiest town of Ionia at the centre of major trade routes by land and sea. A little later in the same century we hear of Pythagoras, who came from Samos, an island just north of Miletos. But what these men did is difficult for us now to ascertain. It is probably safe to concede that they existed, but the specific discoveries attributed to them in later sources seem improbable at best, and palpably legendary in many cases. However, even legends can be informative, as they tell us what people of later generations wanted to hear. There are two aspects worthy of note here: the transformation of geometrical study into 'a scheme of liberal education'; and the mention of Thales disclosing 'principles . . . to his successors'.

The first aspect is significant because one of the most notable aspects of Greek mathematics, by comparison with that of earlier (or, indeed, later) times, was the extent to which it was consciously a different enterprise from that of calculation for practical purposes. The kind of doubts which arose in our evaluation of Egyptian mathematics, over the extent to which it was anything more than practical computation, cannot arise with respect to Greek mathematics. This is because we learn quite explicitly what mathematics is (for them) and what it is for: it is connected—at least in the influential work of Plato—with a high educational ideal, quite divorced from problems of everyday calculation and mensuration. We shall learn more about this when studying Plato's views later, in Section 2.3.

Secondly, we learn, from the mention of 'successors' and Proclus' catalogue of names, of something else of which we have no record in earlier cultures. This is what might be called a *research tradition*, whereby people saw themselves as studying mathematics and developing it, solving certain problems arising in the work of their teachers and predecessors and perhaps creating fresh ones. We must go beyond the writings of Proclus to establish this historically, of course.

We can analyse in several aspects the geometric research tradition as it was developing in the later fifth century BC, just before Plato's time. Any such tradition comprises:

(i) a changing set of *problems*, changing as some are solved or give rise to others;

(ii) particular styles of *method* of approaching those problems;

(iii) means of *justifying* or explaining the solutions of the problems.

We look briefly at how these aspects are exemplified in the work of Hippocrates, who came from the island of Chios (between Samos and Lesbos, off the Ionian coast), and lived in Athens in the latter half of the fifth century: 450–430 BC, perhaps. Proclus refers to Hippocrates as having 'discovered the quadrature of the lune'. A *lune* is, as the name suggests, a moon-shaped crescent: more technically, the shape lying between two intersecting circles. To find the *quadrature* of the lune is to find a square the same size: preferably in a fairly general way, that would be applicable to all lunes, of different shapes and sizes. It is probably safe to infer (unless we have reason to suppose that lunes were an especially significant problem) that this is but one example of a class of problems of trying to find the quadratures of all kinds of shapes.

We are fortunate to have a full description of Hippocrates' work in respect of lunes. This appears in a copy made by Simplicius of an extract from the now lost *History of Geometry* by Eudemus. Eudemus—whose work Proclus probably also drew

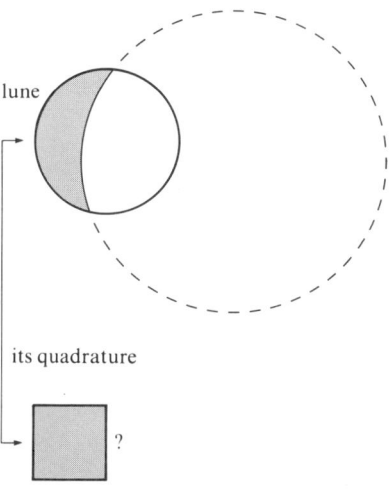

Figure 3

on—was writing in the later fourth century BC, about a century after Hippocrates. This passage is in the Source Book (**SB** 2.B1); please *look at it now*.

Question 3 Read paragraphs 1–2 of this extract and formulate answers to these questions.

(i) In what is described here, does Hippocrates deal with the problem of lunes in general terms?

(ii) What kind of *method*, in broad terms, does Hippocrates seem to be following to reach his conclusion?

(iii) By what means is the solution *justified*?

Study Note If you find this quite hard, it may be because the unfamiliarity (as yet) of the geometric language is distracting you: in that case go on and read the comment. But you do not need to understand the details to answer the questions. One purpose of this exercise is to help you acquire the skill of evaluating a mathematical argument in general terms, without getting too embroiled in detail.

Comment

(i) No, he is looking at the particular case of a lune whose outer edge is a semi-circle. (Indeed, a particular case of that: one whose inner edge is 'similar' to the outer arcs formed in the course of his construction.)

(ii) I think the key word to describe his method is the one I have just used, namely *construction*. I hope you managed to see, even if the terminology was unfamiliar, that in paragraph 2 Hippocrates is described as *doing* something ('circumscribing') to produce a figure—in much the same spirit as Socrates drew squares in the sand for the slave boy. He then applied an argument to show how this led to what was wanted. (See Box 1 for further details.)

(iii) The first paragraph's mention of 'starting-point . . . theorems . . . this he proved . . .' makes it clear that the justification is a logical one, and something more sophisticated than the kind of justification in the *Meno* passage you read earlier. The truth of Hippocrates' claim about the quadrature of the lune depends upon whether 'similar segments of circles have the same ratios as the squares on their bases'. (It does not matter, for our immediate purposes, what this means, but see Box 1 if you are interested.) This, in turn, depends on another result, that 'the squares on the diameters have the same ratios as the circles'. The construction is then devised in such a way as to ensure that these proved or known truths can be applied to the constructed situation. ■

We have no other account as detailed as this of mathematics from the later fifth century BC. But you have seen that we can learn quite a lot from it about the level, style and problems of the research tradition during the period. Thus, we seem to have found ourselves in the middle of a geometrical research programme, concerned with properties of figures bounded by curved lines, using methods of construction and logical justification. Assuming our source to be reliable, this is interesting and rather remarkable. We leave it here for now (returning in *Units 3* and *4* to treat the development of these concerns more fully) and move on to the major figure born at around the time Hippocrates was doing his work: Plato.

Plato was born in Athens in about 427 BC. He is of great significance to our story, on several counts. First, as Proclus said, he 'greatly advanced mathematics in general and geometry in particular because of his zeal for these studies'. Plato was not a practising mathematician especially, and the precise nature of his contribution is still not fully agreed upon, but he seems to have been an important influence upon the mathematicians of his time, by inspiration and direction. Secondly, the works of Plato are our fullest and best source of information about the mathematical developments at that time. He generally introduced mathematics into his discussion illustratively or metaphorically, rather than writing mathematical treatises as such, and the meaning and significance of some of his remarks still give rise to lively controversy. Thirdly, the historical significance of Plato's thought on succeeding generations is almost unparalleled in our culture; it is barely an exaggeration to say, with A. N. Whitehead, that Plato 'defined the complex of general ideas forming the imperishable origin of Western thought'.

Figure 4 Plato

A. N. Whitehead, *Adventures of Ideas* (Cambridge University Press, 1933) p. 132.

Box 1 A note on Hippocrates' quadrature of the lune

We are studying Hippocrates at this point in order to see the general approach of this early example of Greek geometry, not the specific details. However, you may want to check how his proof works, in which case we outline it here.

(We draw attention to some results which Hippocrates must have been able to prove or otherwise believed true.)

Hippocrates constructs the lune as follows. Starting from a right-angled triangle whose two shortest sides are equal, he treats the hypotenuse as the diameter of a circle, i.e. draws a (semi-)circle going through the three corners of the triangle.

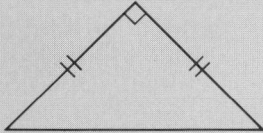
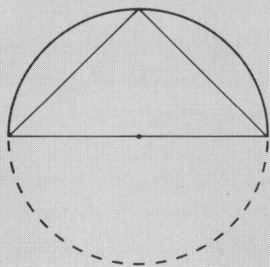

(This requires him to know that a circle centred at the midpoint of the hypotenuse of such a right-angled triangle does go through all three corners.)

On the hypotenuse/diameter he then constructs the inner edge of his lune, a circular arc *similar* to the two circular arcs already forming the outer edge of the lune. *Similar* means that the angles where the curves meet the straight lines are equal.

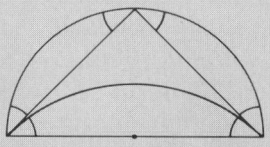

(Again, this requires him to know that a circular arc can be constructed with this property.)

This ends the construction, the rest is reasoning based on the resulting figure. Now he wants to show that the large segment is the sum of the two smaller segments, because it then follows that the lune he has constructed is equal to the triangle he started off with. This will accomplish the quadrature because the triangle can be changed into an equal square without difficulty. But this equality of segments follows from 'his starting-point': 'that similar segments of circles have the same ratios as the squares on their bases'.

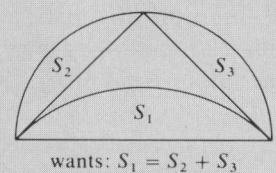

wants: $S_1 = S_2 + S_3$

If the segments are similar (i.e. marked angles equal), then

$\text{seg}_1 : \text{seg}_2 = \text{sq. on base}_1 : \text{sq. on base}_2$

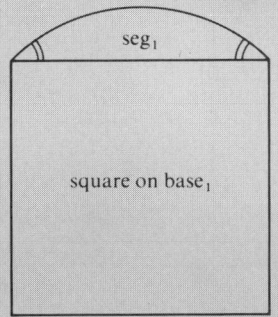
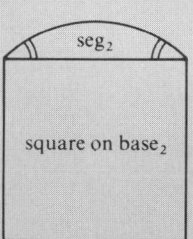

For in this case we have by 'Pythagoras' Theorem' (another result that Hippocrates must have known) that the square on the hypotenuse of the triangle is equal to the sum of the squares on the other two sides. *Therefore*, because of the above equality of ratios, the *segment* on the hypotenuse is equal to the sum of the *segments* on the other two sides.

The prevailing conception of what mathematics is about, from Greek times until now, is Plato's; we look at this in the next section. Much of Plato's influence upon his contemporaries was through his founding a school in about 387 BC, in a part of Athens called Academy (from which all subsequent educational uses of this word are derived). Here he lectured, wrote and directed studies for most of the remaining forty years of his life, dying around 347 BC. Nearly all the fourth-century mathematicians we know of were connected either as students or as teachers with the Academy, as you saw in Proclus' account.

Strictly, there is record only of one lecture delivered by Plato, on *The Good*; see **SB** 2.E8.

Question 4 Read again paragraph 2 of Proclus' summary (**SB** 2.A1), to refresh your memory of the kind of things he claimed for Plato's influence.

Bearing in mind what you have learnt of Proclus and his sources, do you think his account is likely to be fairly accurate in this respect, or more likely to be biased or fanciful?

Comment

The fact that Proclus himself lived so long after Plato must make us a little wary; and as he was in his day the head of the Neoplatonic Academy, this could account for the emphasis he places on Plato. It would not be surprising if he exaggerated a little in a spirit of retrospective pride.

On the other hand, there are reasons for thinking that Proclus' source for this period was Eudemus, who must have been writing within a few years of Plato's death. Also, you have seen that Proclus appears to handle his sources quite carefully in the subsequent paragraph (about Euclid). So, on balance, Proclus' account of Plato and his contemporaries seems likely to be fairly reliable. ■

We shall learn more later about some of the mathematicians associated with Plato. For now, it is important to retain the sense of the research tradition in geometry being especially stimulated and developed by people working in Athens, with Plato's Academy becoming latterly a focal point for such studies. In 367 BC, some 20 years after its founding, the seventeen-year-old Aristotle arrived at the Academy as a student.

The influence of Aristotle (384–322 BC), our other main contemporary source for fourth-century mathematics, has been scarcely less than Plato's. In many respects Aristotle was more pragmatic and down-to-earth than was Plato—and he certainly lacked Plato's literary elegance. Aristotle was especially interested in logical questions. His analysis of logical arguments essentially defined and constituted the science of logic for over two thousand years, and his view of the universe's structure and operating principles lasted (in Western thought) nearly as long. As with Plato, he wrote no explicitly mathematical work as such, but referred to contemporary mathematics in the course of illustrating points of discussion, especially its logical aspects.

Figure 5 Aristotle

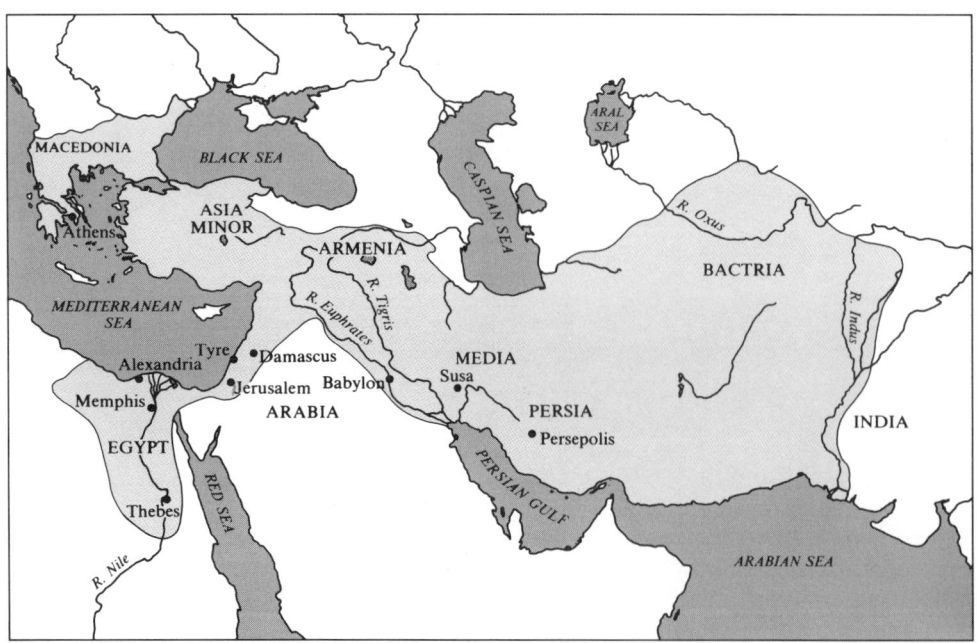

Figure 6 The empire of Alexander

Aristotle was associated with the Academy until Plato's death in 347 BC, and then left Athens for a while. In 342 BC he became tutor to the young prince Alexander of Macedon, who is also important for our story. Alexander became king of Macedon after his father Philip's assassination in 336 BC, and by the time of his own death, only thirteen years later, he had conquered and assembled an empire from Egypt to the Black Sea and from Italy to the Indus. His conquest of the Persian empire brought with it the ancient Mesopotamian cities of Babylon and Susa, as well as the Persian capital Persepolis, and he pushed beyond the old Persian frontiers to the edge of the Indian subcontinent, the site of the equally ancient Indus valley culture. Thus the possibility existed for cultural contacts and the spread of knowledge between the Mediterranean and India, reinforcing earlier trade routes.

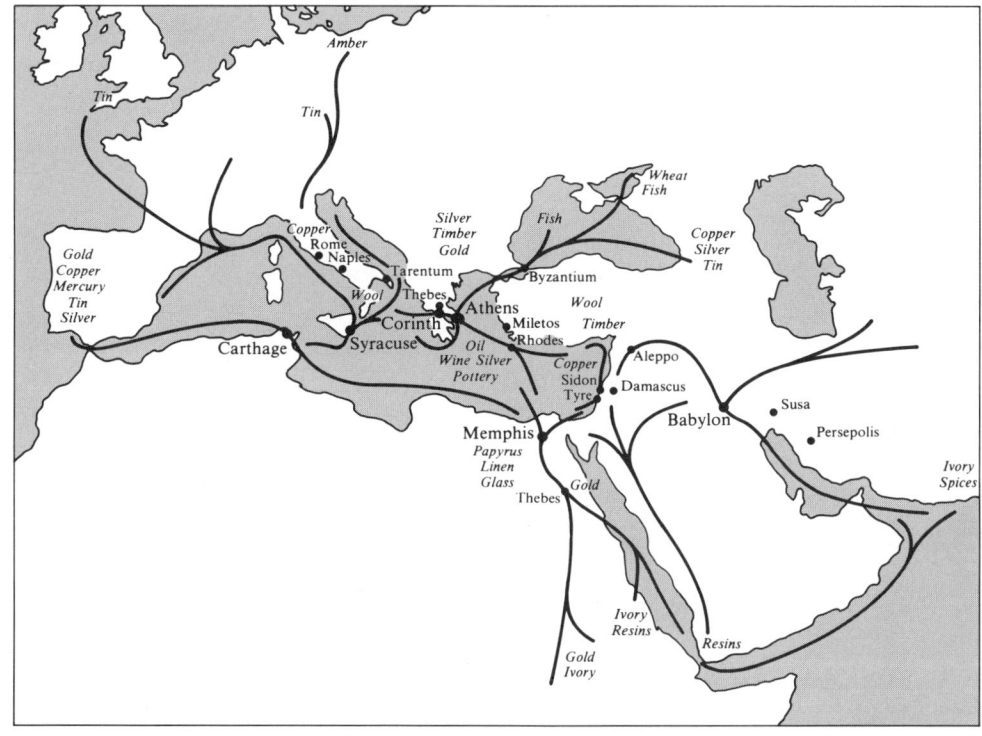

Figure 7 Trade routes of the fourth century BC

Aristotle, meanwhile, returned to Athens and founded a school, the Lyceum. More empirically scientific than the Academy, the Lyceum was supplied by Alexander (according to later sources) both with money and with scientific specimens from his expeditions. For, somewhat like Napoleon two millennia later, Alexander's conquests were attended by a scholarly entourage of engineers, geographers, historians, philosophers and naturalists, and his spirit of intellectual inquiry lasted rather longer than did his empire. In Egypt, Alexander had founded the city of Alexandria. After his death, of malaria in Babylon (in 323 BC), his empire became partitioned among three of his generals, one of whom, Ptolemy, became ruler of Egypt.

Although the smallest third of Alexander's empire, Egypt was also the richest and most easily governed, and Ptolemy's shrewdness may be further gauged by his securing possession of Alexander's body and bringing it to Alexandria, where he set about planning and constructing the new capital city of Egypt. He did this with such success, furthered by his son Ptolemy II who succeeded him in 285 BC, that Alexandria became the cultural centre of the Greek-speaking world for several centuries.

After the third century BC, both the state of our sources, and the kind of mathematics done (where we know about it), move into a different phase, so this seems a good time to pause and consider the nature of the subject matter. In this section, the words *mathematics* and *geometry* have been used somewhat interchangeably. Historically this is not unfair for the period between Plato and Euclid, but you will be able to assess that better for yourself once we have looked more generally at the place of mathematics in Hellenic culture.

2.3 GREEK MATHEMATICS: CLASSIFICATIONS AND LEVELS

The very word *mathematics* is Greek. But it did not take on something like its present meaning consistently until about the time of Aristotle. Previously it had meant something more like 'any subject of study'; so this shift of meaning itself tells us about how our subject was regarded, namely as the most important. Most of our evidence upon the matter comes from Plato, who is not necessarily typical of all Greeks, or even all Athenians. Bearing this reservation in mind, however, we shall look at his views in more detail as representing an informed and articulate contemporary judgement. In his late dialogue *Laws*, a character called the Athenian Stranger—usually taken to represent Plato himself—discusses mathematical education. Please *read that passage now* (**SB** 2.E6) and then answer the following question.

Question 5

(i) Is mathematics to be learned by everyone?

(ii) What is it to be studied for?

(iii) Should those who study it all do so to equal depth?

Comment

(i) No. He is concerned only with what 'the freeborn' should study (which makes the performance of the slave boy in *Meno* all the more striking), and, it seems, only freeborn boys.

(ii) Basic arithmetic and calculations are useful for military purposes and in household management, and also to make the boys 'more awake'. (This is a justification frequently adopted by later educationalists.) The purpose of studying measurement of geometrical objects is to free people from 'ridiculous and shameful ignorance', the nature of which he then goes on to explain.

(iii) Plato says that for the masses, a mathematical education on a par with what is taught in Egypt is sufficient, but to go into it more thoroughly and accurately is 'for a select few'. It does seem, though, that he has more than elementary calculations in mind for the masses, whose studies should advance at least to the point where they redeem the shameful disgrace of not understanding about *incommensurability*. ■

The language Plato uses in relation to this last point is really quite strong. He seems to take it as sub-human, or at least un-Greek, not to know about this geometrical fact. His agitation should give us pause, not to dismiss it as the petulant irritation of a weary teacher, but as evidence of how, for Plato, the concept of mathematical knowledge and truth was very close to a defining characteristic of what distinguished human beings from animals (or from non-Greeks). Plato's exalted vision has not always been shared by subsequent generations, for better or worse. The actual point at issue here is one which occurred in your earlier reading of *Meno*: not all lines are commensurable; you cannot always find a small line that will go a whole number of times into two others which you are trying to compare. The side and the diagonal of a square give an example, as Socrates hinted to the slave boy.

This is discussed further, and proved, at the end of this section

Glance again at the division of mathematical subjects at the start of the *Laws* extract into arithmetic, geometry, and (as a slightly unexpected third) astronomy. Elsewhere, Plato gives a four- or five-part classification, which we shall look at shortly. But first we should reflect on the fact that we have learnt from this passage of Plato about a 'two-dimensional' division of mathematical studies. One dimension is a division into three (or however many) subject areas: cutting across that is a division into knowledge appropriate for the masses and that to be studied by the select few. Just how this latter division is made is not yet clear—or rather, not quite plausible yet—for Plato's image of old men in Athenian coffee-shops setting each other problems concerning incommensurability as readily as they play draughts is perhaps an optimistic one. At the start of the passage the Athenian Stranger seems

The word *subjects*, by the way, is a translation of the Greek word *mathemata*, which illustrates further the point made at the beginning of this section.

to suggest that the difference between mass and select studies is just one of thoroughness and accuracy, somewhat as though the masses would be satisfied with an inaccurate quadrature of the lune (say), whereas the elite like Hippocrates would construct it more precisely. But the distinction must go further, into method and justification and problems—in short, all aspects of the research tradition as defined in the previous section. It is an important historical fact that the first explicit differentiation between two kinds or levels of mathematical studies is found in Plato's analysis. To understand this better, let us look at a passage from Plato's dialogue *Philebus*, where the differentiation is made more explicit and explained. Please *read now* the relevant passage (**SB** 2.E4).

Question 6 Why did Socrates make the distinction between two arithmetics, and between two geometries?

Comment
He explains that it is to investigate whether one kind of knowledge is purer than another, and Protarchus concludes that it is the arithmetic and geometry of the philosophers that are 'immensely superior in point of exactness and truth'. So the calculation and measurement involved in everyday practical activities is sharply distinguished from that of the philosopher (which we may take to be that studied by the 'select few' in *Laws*). In making the distinction Plato uses words such as precision, exactness, purity and truth. ■

As we see in the next unit, mathematical truth and what one must do to establish it were of the greatest concern to Greek mathematicians from the mid-fifth century onwards. So although our source for the explicit description of the distinction between everyday and 'higher' mathematics is from Plato's works, there is no doubt that he was analysing a real and profound difference, which started in Greek times and has persevered ever since.

So, alongside the high geometric research tradition, which we glimpsed in the previous section, there ran an undercurrent of everyday computational mathematics. What did it look like? Ironically, we know rather less about mathematics at this level, practised presumably by most of the citizens, than we know about the high mathematics of the 'select few'; for, on the whole, the former did not write books about it. Some things are clear, however. Greek numerals are of interest as, following an earlier Semitic principle, letters of the alphabet were used to denote numbers. It was quite possible to carry out multiplications (in a way not dissimilar to ours) with these numerals. For a long time historians believed that Greek alphabetical numerals were unsuited to calculation, and must have been used just to record the result of calculations done on an abacus or pebble counting-board (which we know the Greeks also used), until, at the turn of this century, the historian Paul Tannery took the trouble to learn how to use alphabetical numerals and found that computation with them was quite feasible.

To assess the uses to which calculations were put would amount to a survey of practical mathematics within ancient civilisations, and we cannot do anything like justice to that here. But you have seen several pointers already, in what Plato said: in *Philebus* he mentioned navigation, military science, building, agriculture (cows) and carpentry; and in *Laws*, household management and distributions of apples and competitors. Two other aspects of the latter passage are worthy of note: his mention of Egypt as an exemplar of arithmetic teaching, and the problems he gives as illustrations.

The question of connections and the spread of knowledge between ancient cultures—especially between Egypt, Mesopotamia and Greece—is one that has long taxed historians. While the evidence is not as full as we could wish, a spreading knowledge of different computational practices—along trade routes, for instance—seems not implausible. A later Hellenistic commentator spoke of 'the so-called Greek and Egyptian methods in multiplications and divisions, and the additions and subtractions of fractions', showing that the Egyptian computational techniques studied in the last unit were known about in later Greek culture. Again, the Babylonian sexagesimal system was used (with Greek numerals) in later Greek astronomical calculations, though when and how much other Babylonian mathematical lore filtered through to the Greeks is not known for certain.

The particular problems that Plato mentioned in *Philebus* as suitable for teaching purposes (distribution of apples, and something to do with bowls and different metals) are significant, both because they sound not unlike some early Egyptian problems (such as the distribution of loaves, Rhind Papyrus Problem 40) and because later sources contain similar problems. For instance, the scholar Metrodoros in the sixth century AD collected together some ancient and modern arithmetical epigrams, which concern problems such as dividing apples or nuts among a group of people in a particular way, and about finding the weights of bowls. There were also problems that may be familiar to you from your own school-days: if a pipe fills a vessel in one day, another pipe takes two days to fill it and a third three days, how long will all three running together take to fill the vessel? So we can see that there is a long tradition, extending probably much further back than the sophisticated geometric research tradition of the select few, of elementary problems of an educational or recreational nature with some practical everyday overtones—a kind of 'mathematics of the people'.

> See the excerpt from the Greek Anthology **SB** 5.C4.

Finally in discussing the high and the low mathematical traditions, we should note that although the distinction is quite clear as made by Plato, it is more blurred at other times. In fact it may be difficult to 'place' some work which seemed to contain both elements, or which arose within the popular tradition and led to developments in the high research tradition. We shall encounter such difficulties in trying to place the work of Diophantus of Alexandria, in the next section. As further illustration of this point, you may like to look in your Source Book at the 'Cattle Problem' attributed to Archimedes (**SB** 5.C3), which is akin in spirit to the problems you still find in the 'Maths Problem Corner' of the Sunday papers, but of a complexity, despite its innocent wording, quite in keeping with the reputation of Archimedes.

Let us now return to Plato's classification of the mathematical sciences. This is found in its fullest description and explanation in *The Republic*, the great work about his ideal state, written in about 380 BC. There Plato described the mathematical disciplines appropriate for the education of those who are to rule the state: arithmetic, plane geometry, solid (three-dimensional) geometry, astronomy and harmony. Please *read the passage now* (**SB** 2.E2), as far as you feel you need to in order to answer the question.

Question 7 Into which of the two levels of mathematics discussed earlier do these studies fall? Are Glaucon and Socrates agreed about this?

> **SB** 2.E2 is an important and influential passage, well worth your reading right through at some time, but it is quite long, so take your own decision about how much of it you wish to read at the moment.

Comment
It is clear that these subjects are to be treated in their highest, purest form, from Plato's description of what this education is intended to achieve, namely 'minds are to be drawn from the world of change to reality', and 'conversion of the soul from the world of becoming to that of reality and truth'. Even the apparently empirical sciences of astronomy and (musical) harmony turn out in this conception to be quite separate from the physical world; they are a matter for the mind, not the senses.

Glaucon always ends up by agreeing with Socrates, of course, but it is noticeable that he has an ineradicably different conception of the 'use' of mathematics. When each new topic arises, Glaucon cites its everyday, practical use, which is irrelevant to the high educational needs which Socrates has in mind—indeed, it is not even the same subject. In the case of astronomy, for instance, Glaucon takes it to be something to do with telling the seasons, whereas Socrates later explains that it has to do with pure numbers and perfect figures 'perceptible to reason and thought but not visible to the eye'. This amusing dramatic device enabled Plato to make his distinction both clearly and effectively. ■

One historical significance of this passage is that from it stemmed the *quadrivium*, the advanced part of mediaeval higher education (from about AD 1000 onwards). This consisted of arithmetic, geometry, music and astronomy, and was preceded by the *trivium* of grammar, logic and rhetoric. The actual content of the mediaeval studies sometimes had little to do with Plato's conception—being probably rather closer to Glaucon's in places—but the categorisation, and the placing of mathematical studies at the summit of the liberal arts course, derives from Plato through various later Greek and Latin writers and educationalists.

> The splitting up of geometry into plane and solid (three-dimensional) was not taken up—indeed the discussion of solid geometry seems addressed by Plato to his contemporary Athenians.

The division of mathematical studies probably did not originate with Plato, but developed during the previous century. For there is one other contemporary reference to it, in a fragment by Plato's friend Archytas. Archytas lived in Tarentum (Taranto, in Southern Italy), and was referred to by Proclus as one of those 'by whom the theorems were increased in number and brought into a more scientific arrangement'. Archytas wrote:

> I think that those concerned with the sciences [*mathemata*] are men of discernment, and it is not strange that they should think correctly about the nature of particular things. . . . And so they have handed down to us clear knowledge of the speed of the heavenly bodies and their risings and settings, of geometry, numbers and, not least, of the science of music. For these sciences [*mathemata*] seem to be related: they are concerned with the first two kinds of what is, which are related.

SB 2.D1

Archytas' explanation of why these subjects are related is perhaps not very clear, but he states plainly that the four quadrivium subjects have been 'handed down to us'. According to later sources, this was by the Pythagoreans, that is, followers of Pythagoras. Indeed, Archytas is always spoken of as a member of the Pythagorean brotherhood, so it may well have been the case.

At the end of the *Republic* extract, Plato made the similar point to Archytas, that all these studies are related and have 'kinship and common ground'. How they do so is rather clearer in the way Plato has been explaining the nature of the subjects. We shall not however pursue astronomy and harmony any further here. We must now fill in a gap that you may have begun to feel. You know what Plato means by *geometry*. The *Meno* extract that we started with is a paradigm case, and we looked at the geometric research tradition in the second section. But what of *arithmetic*, the leading study of Plato's classification? The discussion in *Republic* is clearer about what the subject should achieve than about what it is. Or perhaps Plato did tell us, but what he means is too unfamiliar for the twentieth-century reader to recognise? Let us look again at that passage.

Question 8   Re-read the section on arithmetic (in **SB** 2.E2). What would you say arithmetic is *about*? Can you find anything which seems to amount, however obscurely, to mathematical content in the passage?

Comment
Arithmetic seems to be about the nature of numbers; in particular, there is a contrast between unity and plurality. This is still hard to pin down, but we are given what looks as though it ought to give a substantial clue when Socrates describes an argument involving 'the experts'. We learn that the unit has two properties: all units are the same, and they cannot be divided into parts. We also learn that if you do try to divide the unit, the experts reply by multiplying it. That is a little obscure, but we seem to be making progress. I would gloss the argument by observing that it makes sense if, when you try to divide the unit (into halves, say), the expert redefines the unit so that your half is the new unit, so the old unit is now double what it was. However this may be, it seems that units and their multiples (that is, plurality) are at the kernel of Plato's concept of arithmetic. ■

For further clues about the nature of arithmetic, we need to search elsewhere. In fact, three of the books of Euclid's *Elements* are devoted to what we have just been teasing out of Plato. Please *look now* at the definitions in *Elements* VII (**SB** 3.D1) to see the kind of thing that happens there.

There are several interesting things to notice about these definitions.
(i)   They start off by defining a *unit*;
(ii) then a *number* is defined in terms of that;
(iii) in such a way that a unit is not a number (because a unit is not a multitude composed of units);
(iv) then three relationships between numbers are defined: *part of*, *parts* (*of*) and *multiple of*;

(v) then various sorts of numbers are defined: *even, odd, even-times even, even-times odd, odd-times odd, prime, composite, plane, solid, square, cube, perfect*;

(vi) and a few other definitions, of which we shall just note Definition 20, which specifies when *four* numbers are *proportional*.

(vii) So some of these definitions are applicable to a single number considered by itself (the definitions in (v) above); some are relationships between two numbers considered jointly ((iv) above); and for Definition 20 the relationship between four numbers needs to be considered.

Now, this is probably half-familiar to you: for you know terms like *even, odd, prime, square, multiple of*. Yet Definitions 1 and 2 are rather strange. There is one feature in particular that we might not have noticed, had not Plato singled it out for mention: the unit is not something you can divide, so all the arithmetic built up through these definitions is an *arithmetic of whole numbers*—there is not a fraction in sight.

This is an important point, so it is worth explaining the implications with an example. In the development of this arithmetic, that is, seeing what follows from these definitions, things happen which *we* would treat using fractions, and indeed we can barely understand otherwise. For instance, *Elements* VII, 6 proves that

VII, 6 means Book VII, *Proposition 6*.

> If a number be parts of a number, and another be the same parts of another, the sum will also be the same parts of the sum that the one is of the one

which plays a similar role in this arithmetic to the fractional manipulation

$$\text{if} \quad a/b = c/d, \quad \text{then} \quad (a + c)/(b + d) = a/b$$

in ours. But although it helps our mathematical understanding to make this analogy, we must be careful not to attribute fractions like ours to Greek unit-arithmetic, which would devastate precisely the distinction that Plato was at such pains to make.

However, you should not think that studying unit-arithmetic is a constant struggle against temptation. In Euclid's presentation of it there are some beautiful results which have been a part of mathematics ever since. We look at just one of these, to give you the flavour of Euclidean number theory. This is *Elements* IX, 20, showing that there is no limit to the number of primes. Please *read it now* (**SB** 3.D2). Try to spend as long studying it as you need and can afford, for you should find it rewarding. This proof has been considered ever since as one of the most beautiful and elegant in all mathematics.

The English mathematician G. H. Hardy (1877–1947) said that it is 'as fresh and significant as when it was discovered—two thousand years have not written a wrinkle on [it].' (*A Mathematician's Apology* (Cambridge University Press, 1967) p. 92.)

Question 9

(i) One of the aspects most admired about this proof is its *structure*. Try to summarise this—pick out the structural framework.
(ii) The proof shows only that if there are three prime numbers, there must be a fourth. In view of this, do you feel that Euclid has proved the general claim made in the proposition?

Comment

(i) The overall strategy is revealed in the first two lines, following directly from the statement of what is to be proved: 'Prime numbers are more than any assigned multitude of prime numbers.' He assigns a multitude (*A*, *B* and *C*) and will show that there are more prime numbers than this.

First he constructs another number ('*EF*'), which is effectively *A* times *B* times *C*, plus one; and then argues as in Figure 8.

(ii) There seems to be an assumption which Euclid does not spell out, that for any other assigned multitude of primes the proof would be exactly analogous. Or to put it another way, that the case of three primes adequately represents the general case. Granting this assumption, the proof is indeed perfectly general. ∎

There is one further historical aspect to examine. In order to elucidate Plato's conception of arithmetic, we leapt forward nearly a century, from c. 380 to c. 300 BC. Was that justifiable? Did nothing happen in between?

We have no evidence of very much happening in this area during the fourth century BC, and it is generally supposed that Euclid was setting down the arithmetical theory of a century or so before his time, though the extent to which he may have

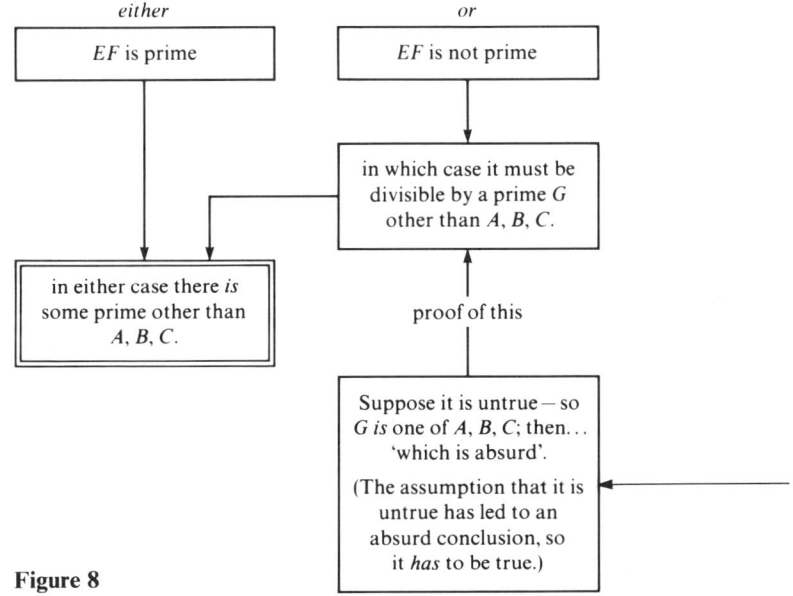

Figure 8

Notice what happens in this box. A statement is proved by showing that its opposite leads to an absurdity or contradiction (here, that the unit could be divided). This kind of proof is called *reductio ad absurdum*; we shall see other examples later.

formalised and tidied it up into a logical structuring is still debated among historians. It seems that the geometric research tradition was more active and exciting, with more interesting problems, from at least the time of Plato onwards. For one of the aspects of a research tradition is that it is fuelled and sustained by new and unsolved problems; once the problems are solved, people will turn their energies to other areas. So it could well be that the nice, self-contained, coherent arithmetical theory of Euclid's *Elements* VII–IX said all that there was to be said on the subject (at the time). It was, indeed, so 'pure' as not to be applicable to other mathematical studies, which might otherwise have been a source of fruitful problems for development. You saw in *Meno*, for instance, how the slave boy's initial attempts to solve a geometrical problem arithmetically led nowhere, so Socrates diverted his attention into considering a geometrical solution.

The inapplicability of unit-arithmetic to some geometric figures seems to have become known towards the end of the fifth century BC, perhaps during Plato's youth. It is discussed by Plato in several places—his diatribe against 'shameful ignorance' in the *Laws* is related to this—and is one of Aristotle's favourite examples. Let us look at what we can learn from Aristotle. Please *read now the short extract* **SB** 2.H6(b).

As is usual with Aristotle, he is here using a mathematical example to illustrate a more general point; presumably the mathematics would have been familiar to his original hearers, as his treatment is rather allusive. You may well have recognised his general logical point—he is discussing the argument structure we noticed above, *reductio ad absurdum*, where in Aristotle's words you 'prove the original conclusion hypothetically when something impossible follows from a contradictory assumption'. Hence the overall structure of the mathematical proof he is referring to is clear: if we assume that the diagonal of a square is commensurable with its side—that is, that unit-arithmetic is applicable through finding a unit length measuring both lines—then an absurdity arises, involving the conclusion that 'odd numbers are equal to even'. Hence the initial assumption must be wrong, so the diagonal is not commensurable with the side, and unit-arithmetic is inapplicable.

Various reconstructions have been made by historians of the details of this proof; we give the most time-hallowed one.

Suppose the diagonal AC and the side AB are commensurable. Then there exists a unit length in terms of which AC is n units and AB is m units, and n and m are the smallest possible. This means that the numbers n and m have no divisors in common (other than the unit). Since the square on the diagonal AC is twice the square on the side AB, as we saw in the *Meno* passage, we have $n^2 = 2m^2$.

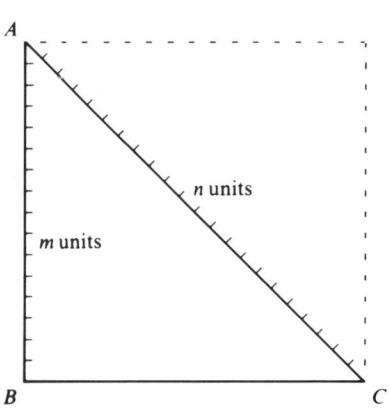

Figure 9

It follows that n^2 is an *even* number (since it is *twice* some number)—so n is an even number; and so m *must* be an *odd* number (because n and m have no divisors in common). Since n is even, $n = 2p$ (say), and its square being equal to $2m^2$ gives us $(2p)^2 = 2m^2$; so $4p^2 = 2m^2$; so $2p^2 = m^2$. Thus m^2, and so m, now turns out to be an *even* number. But we already know that m must be an odd number. So we have shown that m is both an odd number and an even number, which is absurd.

If n^2 is even then n *must* be even, because the square of any odd number is odd.

It is not known whether this was the original proof of the result that the side and the diagonal of the square are incommensurable. Nor is it known whether the case of the square was the first in which the phenomenon of incommensurability was recognised. (Some historians have argued that this recognition took place in connection with investigations of the regular pentagon, whose side and diagonal are also incommensurable.) It is not important to resolve this for the purposes of our story. But notice one significant aspect of the result you have just seen proved: assuming you found it convincing, and now believe the result, you do so *only* because of the *proof*. The result has very little plausibility without proof accompanying it. This is an entirely new situation. Other results earlier in the unit—Hippocrates' quadrature of lunes, say, or the unlimited number of primes—had proofs which acted more so as to corroborate what might have seemed quite likely beforehand. But the discovery that two lines were incommensurable, and the proof, must have been more-or-less simultaneous. Indeed we might go further and say that its first proof must have constituted its discovery, though the details of this event are no longer known.

A consequence of this observation is that the phenomenon of incommensurability could only have come to light in a culture employing some notion of logical proof. We can illustrate this by considering the contrast between the above result and an interesting Old Babylonian tablet (**SB** 1.E1(g)), which you might like to study briefly now.

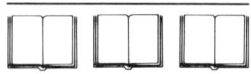

The tablet gives a very good approximation to the length of the diagonal of a square. It is an impressive tribute to Babylonian computational skills. But in the light of our present discussion, can we infer anything more about the scribe's intentions or state of mind? It is reasonable to suppose he knew that the number written down for the diagonal is not exact, but did he know that no number of sexagesimal places, however far continued, would give an exact measure of the diagonal in relation to the side? There is no evidence he did, and indeed by the previous argument it seems unlikely. The mathematical context in which an awareness of incommensurability is possible involves a concept of proof which is alien to what we know of Babylonian mathematical style.

This example illustrates how large a gap there is between the two levels of mathematics emphasised by Plato. It could be that both Plato and Aristotle paid so much attention to incommensurability just because it symbolised, better than any other mathematical result of the later fifth century, the power of the newly emerging techniques of logical justification—a result whose very credibility is inconceivable apart from its proof.

The story of proof is taken up in more detail in the next unit.

2.4 MATHEMATICAL TRADITIONS IN THE HELLENISTIC WORLD

By the beginning of the third century BC, the intellectual focus of the Greek-speaking world was no longer Athens, but Alexandria. This involved more than a mere shift of locale, in that it coincided with a change of intellectual style so marked that historians refer to it by a fresh name, the 'Hellenistic' period. Typical of the new Alexandrian style were two institutions established by Ptolemy, the Museum and the Library. The Museum was a scientific research institute (in our terms), founded in the spirit of Aristotle's Lyceum, which proved extraordinarily productive in the sciences of astronomy, mathematical geography, anatomy and physiology, as well as in mathematics. The Library at Alexandria developed rapidly to become the greatest library of antiquity, containing at its peak perhaps 400 000 or more rolls of papyrus; even by about 250 BC, the library catalogue alone occupied 120 rolls.

Each of the three most famous Greek mathematicians—Euclid, Archimedes and Apollonius—had associations with Alexandria during the third century. As we shall look at their works in the next two units, here we only indicate their chronological placing. Almost nothing is known of Euclid's life, but he may have learnt his mathematics at the Academy in Athens, to judge by the contents of the *Elements* as well as by Proclus' claim that Euclid was a Platonist. That he taught in Alexandria (perhaps in about 300 BC) can be inferred from a remark made by a later commentator, that Apollonius was taught there by Euclid's pupils. Euclid's life probably overlapped that of Archimedes, the greatest and most versatile of Greek mathematicians in the eyes of posterity. Archimedes lived most of his life in Syracuse, in Sicily; indeed, it is not known for certain whether he ever visited Alexandria, but he had friends there and corresponded with mathematicians at the Museum. The lifetime of Archimedes (who died in 212 BC) in turn overlapped that of Apollonius, the 'great geometer' of antiquity. Apollonius studied in Alexandria and travelled somewhat round the eastern Mediterranean, but little else of his life is known. His influential work *Conics* we look at later, in *Unit 4*.

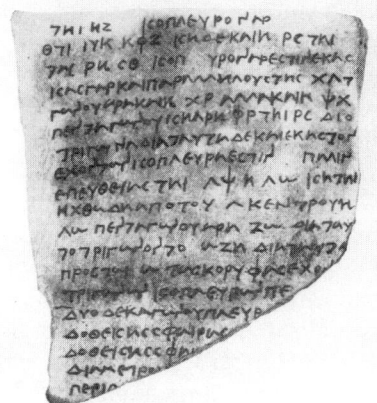

Figure 10 The earliest known physical fragment of Greek mathematical writing, relating to Euclid's *Elements* XIII. It is on a piece of broken pot, found in southern Egypt in 1907, and dates from about 250 BC.

These three mathematicians form, in their different ways, the culmination of the Greek geometric research tradition. Their works were preserved (to a greater extent than were those of their predecessors), but were not significantly advanced, developed or built upon for nearly 2000 years. We take up that story in Block II. But the shift in intellectual style mentioned above as characteristic of the new Hellenistic age is revealed in their other interests. Euclid wrote a book called *Optics* (see **SB** 5.A3(b)); Archimedes had a high reputation in practical and theoretical aspects of mechanics; Apollonius was noted for his astronomical studies. Although from the viewpoint of 'pure' mathematics the later Hellenistic age is on the whole somewhat uneventful, the mathematical sciences in a broader sense were pursued vigorously and to good effect.

In particular, the high Platonic distinction between mathematics and practical computation became rather blurred as time went on, and we find works in which great geometrical sophistication is combined with high computational skill. Nowhere was this more true than in astronomy.

Disregarding Plato's injunction to 'ignore the visible heavens', Alexandrian astronomers developed the mathematical means to account for and predict celestial movements and phenomena, to a high degree of precision. This culminated in the work of Claudius Ptolemy, in about AD 150, called the *Mathematical Synthesis* (more commonly known by its Arabic name, *Almagest*). This work was as effective as Euclid's *Elements* in summarising and making redundant the works of his predecessors, and was the definitive astronomical text for the next 1500 years. We cannot study Ptolemy's work here, or indeed the history of mathematical astronomy, in any detail. It is useful, however, to record the existence of a work of very considerable mathematical skill and expertise, but with different problems and methods from those previously described as characterising Greek research

Claudius Ptolemy has no connection with the ruler Ptolemy whom we mentioned earlier.

geometry. There is certainly considerable Babylonian influence on the work of Ptolemy and his predecessors, not least in their adoption of the sexagesimal fractional system for astronomical computation. Babylonian astronomers had developed a systematic mathematical astronomy from about 500 BC onwards, fully worked out by the time of Alexander the Great's conquests, which made it distantly accessible within the Greek cultural orbit. But Greek mathematical astronomy, as developed from about that time up to Ptolemy four centuries later, is different in principle, with the calculations based on a geometrical model of the planetary movements rather than the more strictly arithmetical procedures of the Babylonians.

One aspect of Ptolemy's work, which provides evidence for the existence of sophisticated computational skills in Alexandria, is his *trigonometry*. To determine numerical relationships between sides and angles of triangles—fundamental to astronomy, mathematical geography, navigation, and so forth—he used the measurements of chords in circles, rather than our sines and cosines (a later Hindu invention), although it amounts to the same thing. Besides the *Almagest*, other works by Ptolemy testify to the range of mathematical sciences of his period: he wrote works entitled *Optics* (on vision, light and the mathematics of reflection and refraction), *Harmonics* (on music theory) and *Geography*. This latter tome was an influential work which was the first to employ latitude and longitude in a systematic way, and in which he laid down two important map projections. (This work contained an underestimate of the size of the Earth which contributed, much later, to Christopher Columbus believing China to be more readily accessible across the Atlantic than turned out to be the case.) Ptolemy also wrote a work on astrology, *Tetrabiblos*, to the dismay of some historians of science: 'It is a great pity' wrote the distinguished historian George Sarton, 'that Ptolemy wrote it' and G. J. Toomer has described it as 'a specious "scientific" justification for crude superstition'. However this may be, the *Tetrabiblos* was for a long time a very influential work on its subject, and for us acts as good evidence of the breadth of Hellenistic mathematical activity.

George Sarton, *Ancient Science and Modern Civilisation* (Harper, 1959) p. 60; G. J. Toomer, 'Ptolemy', in *Dictionary of Scientific Biography*, vol. XI (Charles Scribner's Sons, 1975) p. 198.

Ptolemy also wrote a work on mechanics, which has not survived. Perhaps this loss is an indication that his work in this area was less highly regarded. For an evaluation of the state of mechanics during this period, we turn to a passage by the fourth-century AD writer Pappus of Alexandria. Please *read that now* (**SB** 5.A2).

Question 10 What division of mechanics does Pappus mention? To what extent is this comparable with Plato's division of mathematics into two levels?

Comment
The division of mechanics, which he attributes to 'the mechanicians of Heron's school', is into *theoretical* (sciences) and *manual* (arts). It is tempting to read into this a distinction analogous to Plato's for mathematics, but on balance I would say that they are not quite the same thing. Plato's distinction hinged on greater or lesser approaches to truth, by which criterion the 'lower' level was effectively unnecessary; whereas the Pappus/Heron distinction is more of two complementary and equally necessary aspects of mechanics, to be split between different people for purely pragmatic reasons. ∎

Pappus' discussion of Archimedes in the last paragraph is interesting, not only for what it tells us about the breadth of Archimedes' interests and reputation (five centuries after his death), but also for the slightly nervous justification for applying geometry to everyday life—the final sentence makes no sense unless there were people around who protested, perhaps on Platonic grounds, against the sullying of pure geometry by application to everyday arts and sciences.

To exemplify the practical mathematics tradition in Alexandria, let us consider the work of Heron, who was mentioned by Pappus throughout this passage, and who lived in the century before the astronomer Ptolemy. Heron wrote a large number of works on mechanics, covering all the things described by Pappus in his middle paragraph, but what is of particular interest to us is a work called *Metrica*. This was rediscovered only in 1896 in Constantinople (now Istanbul), in a manuscript of the eleventh or twelfth century, and turns out to be an intriguing blend of practical mensuration and methodological treatment more akin to the pure geometric

We know that Heron lived around AD 62, from his reference to an eclipse which can only have occurred that year.

...esearch tradition. You can see this through examining the extract given in the Source Book (**SB** 5.A5).

Question 11 Read the whole extract through fairly quickly, to enable you to characterise broadly what is going on. Then comment on the following question. Does this passage confirm the assertion made above about *Metrica* being a blend of practical mensuration and pure geometry?

Comment

This passage falls into two parts. The first paragraph gives a sequence of instructions for determining the area of a triangle if you know the lengths of the sides. The rest is the geometrical proof of the formula that is, in effect, being used. The first part is reminiscent of a Babylonian problem text, though Heron says explicitly that there is a general method, for which he gives the successive steps in a particular example. Also the geometrical justification which follows is in the pure geometric tradition—the translator's additions in square brackets show how many results from Euclid's *Elements* are used in the course of the proof.

You may have noticed a feature in the first paragraph that is not paralleled in the later geometric justification. The method consists of two parts: first, reaching some number whose square root will be the answer; secondly, applying an independent set of recipe-like instructions for working out an approximation to the square root if you cannot immediately find an exact one. Heron says that if this gives as accurate an answer as you want, fine; if not, you do it again. Thus we see that the practical use of all this is at the forefront of his attention. ■

You saw earlier that Greeks of the fourth century BC believed, rightly or wrongly, their mathematics to be derived from the Egyptians. In a similar way, Hellenistic thinkers attributed to some distant golden age the things they valued. The early centuries of our era saw, throughout the Graeco–Roman world, a flourishing of many religious cults and philosophical beliefs. One such was what we call *neo-Pythagoreanism*, a revival (as they saw it) of the knowledge and beliefs of Pythagoras. Especially influential was the work of Nicomachus of Gerasa, who probably lived in Judaea around AD 100. His *Introduction to Arithmetic* was a basis for arithmetic teaching in the West for the next 1500 years. You can see from the opening of this work (**SB** 2.D3) his tribute to Pythagoras, leading to an explanation of the 'quadrivium' classification of mathematical studies. We receive a good idea of the neo-Pythagorean approach by comparing Nicomachus' treatment of an arithmetical topic with Euclid's, four centuries earlier. An interesting example is that of *perfect numbers* (*Elements* VII, Def.22).

VII, Def.22 means Book VII, *Definition* 22.

Question 12 Read the discussion by Nicomachus of perfect numbers, and Euclid's *Elements* IX, 36 (both in **SB** 3.D3). Do not work through Euclid's proof in detail (unless you want to), but satisfy yourself of its general characteristics. Note that his proposition has partly the same effect as paragraph 4 of the Nicomachus passage. Then answer the question: what differences are there between the two treatments?

Comment

I noted at least four differences. First, Nicomachus is much easier to follow. If you want to produce an example of a perfect number, he is more 'reader-friendly' than is Euclid: just follow the instructions, like a recipe-book. On the other hand, their enterprises are different in that Euclid *proves* his result, which Nicomachus does not do. Thirdly, and no doubt connected with this lack of proof, Nicomachus makes some rather audacious claims about perfect numbers, which range from the false (that there is one perfect number in the units, one in the tens, one in the hundreds, and so on) to the still unproven (that his method produces every perfect number, and that all perfect numbers are even). Euclid made a more restricted claim, but one capable of proof. Fourthly, Nicomachus' discussion invokes ethical analogies (in paragraph 2), unlike Euclid's. (Of course, the very word 'perfect' has moral connotations, so Nicomachus may well have been reflecting an old tradition or conception.) ■

So the treatments are very different, and it is tempting to see in Nicomachus a debased popularisation of Euclidean number theory, an assimilation into 'low' mathematics of the high classical past. But it is worth noting that the topic of perfect numbers has little to do with everyday computational mathematics; in that respect Euclid and Nicomachus are at one.

Nicomachus was particularly emphatic about his debt to Pythagoras, and certainly his somewhat theological conception of number epitomises what tradition associates with the Pythagoreans, as in this passage:

> All that has by nature with systematic method been arranged in the universe seems both in part and as a whole to have been determined and ordered in accordance with number, by the forethought and the mind of him that created all things: for the pattern was fixed, like a preliminary sketch, by the domination of number preexistent in the mind of the world-creating God.

Nicomachus, *Introduction to Arithmetic*, Book I, ch. 6.

(You may like to compare this with the views that Aristotle attributed to 'the so-called Pythagoreans', **SB** 2.H4.) But many other classical Greek concepts enter Nicomachus' exposition, especially ideas from Plato. For our purposes in gaining an impression of Hellenistic culture, it does not matter precisely what sources Nicomachus drew upon. The thing to notice is that there were several of them, they came from the fairly distant past, and Nicomachus may not have been much better able than we to discriminate between them.

The roving searchlight we have been playing across the Hellenistic scene thus far has come across mathematicians of great versatility and application, but none, from the 'pure' mathematical perspective, of the calibre of the giants of yore, Archimedes and Apollonius, or of their fifth- and fourth-century predecessors. Their fame and works lived on—for instance, you have seen Pappus' tribute to Archimedes—but the high mathematical research tradition seems by this time to have petered out. There is one Alexandrian mathematician, however, of great subsequent influence, whose work is a puzzle for the neat categorisations beloved of historians. He is Diophantus, who probably lived in the middle of the third century AD, somewhere perhaps around AD 250. The puzzle is neatly put by the historian Wilbur Knorr: 'if Diophantus had not existed, no historian of ancient Greek mathematics would have invented him'—in other words, but for the evidence of Diophantus' book *Arithmetica* there would be no reason to suppose that anyone would have been creating a corpus of mathematics like it, at any time throughout the Greek period. You will be looking in more detail at his work later (in *Unit 5*), but it may be interesting to make a preliminary survey here.

W. Knorr, *American Mathematical Monthly* (Feb. 1985) p. 150.

Question 13 Browse through the Diophantus section in the Source Book (**SB** 5.D), and formulate an answer to the following question. Which of the mathematical traditions you have been examining does his work seem to fit into best?

Comment
It is clearly not part of the geometric research tradition. It is not about geometry, and there is none of the formal proposition/proof form one associates with that. On the same grounds, it does not look like a development of Euclidean number theory (*Elements* VII–IX, which you examined in the previous section), even though the name *Arithmetica* might lead one to expect this.

On the other hand, it is also difficult to place the work firmly *within* the practical/computational tradition. In comparison with Heron, for example, Diophantus has no interest in approximate solutions to his problems; and compared with earlier problem traditions (such as the Rhind Papyrus), he is concerned with numbers simply in themselves, not numbers of something-or-other. Also, the problems of Diophantus can be vastly more complicated or subtle than, say, those in the later *Greek Anthology*. Again, it seems to be clearly 'straight' mathematics, carefully thought through and with no apparent neo-Pythagorean overtones. There is little sign of the influence of Nicomachus here. In short, Diophantus' work does not seem to fit in with anything we have seen, although perhaps it would not seem quite so out of place if we were less ignorant of some aspects of early mathematics. ∎

SB 5.C4, mentioned in Section 2.3.

2.5 THE COMMENTATING TRADITION

From the third century AD onwards we see the growth of another tradition, of a different emphasis. It was pre-eminently the age of the *commentator*. The way to disseminate, teach, and indeed *do* mathematics became one of writing commentaries, explanations and expansions upon the work of great mathematicians of the past. Of course, this is not something completely new, but more a shift of emphasis and balance. For even by the time of Aristotle, mathematics *had* a past, insofar as his pupil Eudemus wrote histories of geometry and of arithmetic; and the great landmarks, such as Euclid's *Elements* and Ptolemy's *Almagest*, incorporated and consolidated earlier work. The commentating tradition, in which the prime focus is upon preserving, clarifying and transmitting the mathematics of the past, does nevertheless represent a different way of conceiving mathematical and teaching activity, which continued through various cultures right up to the seventeenth century.

Why did this development take place? To observe merely that there were no longer mathematicians like Hippocrates or Apollonius is a statistical comment rather than an explanation. The answer, if there is one, may lie in the interplay of the mathematical research tradition with the political and social circumstances of the Hellenistic world. One aspect of Greek mathematics to which we have as yet given little thought is how it was taught and transmitted at the time. The fact that what we have, necessarily, are texts and other written sources about Greek mathematics should not lead us to overlook that there was a strong oral and face-to-face component to Greek mathematical communication and development. (Again, Socrates and the slave boy can act as a symbolic paradigm of this.)

You will come to see later that, in the work of Apollonius and even in parts of Euclid's *Elements*, products of the high research tradition can seem almost meaningless without an accompanying explanation. The development of strict logical formulation was to ensure the truth-status of results, not their motivation or understandability. This is all very well so long as a teaching and research community remains in being, to transmit its values, practices and understandings to each fresh generation, but conditions in the Hellenistic world stopped this happening. Even the Academy, of which Proclus became head in the fifth century AD, shared at best a name and distant ideals with what Plato had founded. There is no evidence that the mathematical research tradition of fourth-century BC Athens was sustained and maintained through all the intervening centuries—it had probably lasted but a century or two at best. So, scholars such as Pappus and Proclus were faced, just as we are, with documents from the distant past but little by way of accompanying oral explanation. Hence an important activity became the writing of commentaries, of varying quality, explaining and expanding obscure portions to the best of one's understanding, adding any new result that occurred, citing any other document that seemed to cast light on the matter in hand, and so on.

Pappus lived in Alexandria probably about AD 320 (he speaks of an eclipse of the sun which took place in that year). Not only is he one of our principal sources for earlier Greek mathematics, but as a mathematician in his own right some of his geometrical results were significant and influential much later. Without his study of what remained of the Library of Alexandria in his day, our knowledge and understanding of Greek mathematics would be greatly impoverished. The later history of the Library is obscure. It seems to have fallen into gradual decay, perhaps from as early as the first century BC. It is a frustrating reflection for the historian that once all Greek mathematical learning was collected together in one place; but the ravages of time, Romans and Christians ensured that by the time of the Moslem sack of Alexandria in AD 646 there was very little of the Library left to destroy.

From about the third century AD onwards, a set of beliefs known as *Neoplatonism* became prominent in intellectual circles of Alexandria and elsewhere. As the name suggests, it was thought of as drawing from the truths of Plato's philosophy; but, as with neo-Pythagoreanism, many other things were included. The end result was an idealistic religion of high ethical purity, which inspired many of the best scholars of the age. One aspect of Neoplatonic beliefs relevant to our concerns is that it is a philosophy of revelation. Truth is revealed by the divine, not reachable by human thought alone, and such revelations had been granted to the wise men of old Pythagoras, Plato, Nicomachus . . . Thus we have another slant on the commentating tradition: wanting to establish and understand what had been written in the past was not simply curiosity about past mathematics, but could also be important for philosophical or religious belief.

There are two other Alexandrian commentators we should take notice of. Theon of Alexandria, whose edition of Euclid's *Elements* our present knowledge is essentially based on, lived later in the fourth century (he mentions seeing two eclipses of AD 364). All but one of the extant Greek manuscripts of the *Elements* are copies of the version which he prepared for the use of his students at the Museum (of which he is the last attested member). This is a mixed blessing, as he does not always seem to have understood it very well. His daughter Hypatia is history's first recorded woman mathematician. Her fame has come down to us less for her learned commentaries (which have not survived) on Apollonius' *Conics*, Ptolemy's *Almagest* and Diophantus' *Arithmetica*, than for the striking image provided by the manner of her death in AD 415: the wise pagan philosopher hacked to pieces in a Christian church by the energetic forces of emergent Christianity.

The murder of Hypatia, who was head of the Neoplatonic school in Alexandria, effectively brought the Neoplatonic movement there to an end. The last great Neoplatonic philosopher in Athens was Proclus (c. 410–485). Although our interest in him has been primarily for what we can learn from him about the history of Greek mathematics, his own influence on subsequent thought was considerable. In the words of two historians at the beginning of this century:

> The works of Proclus, as the last testament of Hellenism to the church and the middle ages, exerted an incalculable influence on the next thousand years. They not only formed one of the bridges by which the medieval thinkers got back to Plato and Aristotle; they determined the scientific method of thirty generations, and they partly created and partly nourished the Christian mysticism of the middle ages.

A. Harnack and J. M. Mitchell, 'Neoplatonism' in *Encyclopaedia Britannica* (11th edn, 1910–11).

It is worth bearing in mind that when, later in the course, you meet references to Renaissance Neoplatonism (important in the work of, for instance, Johannes Kepler), this means a set of beliefs deriving ultimately from Plato but as seen through the eyes of Proclus who lived nearly half-way in time between Plato and Renaissance.

We mention one more writer of the period whose philosophy was Neoplatonic and whose work was of great historical influence—the Roman aristocrat Boethius (c. AD 480–525). We have said little about the Romans, as their interest in mathematics seems in general to have been confined to elementary practical mathematics. (One example comes from Vitruvius, first century BC; see **SB** 5.A6.) Boethius, however, appears to have set out to translate into Latin, and comment upon, as much Greek learning as he could gain access to. He was imprisoned for political reasons and then killed before he could put all of his plans into effect, but he wrote an *Arithmetic* based on that of Nicomachus, and a *Music* derived from Nicomachus, Ptolemy and Euclid. It was he who, as far as we know, first used the word *quadrivium* for the classification of the mathematical sciences (see **SB** 2.D4). It is not certain whether he also compiled works on *Geometry* and *Astronomy*, although these were attributed to him in the Middle Ages. That these and other works were given Boethius' name testifies to the fact that his role in transmitting knowledge of the quadrivium to the West caused Boethius to become revered, for a time, as one of the great mathematical giants of the distant past.

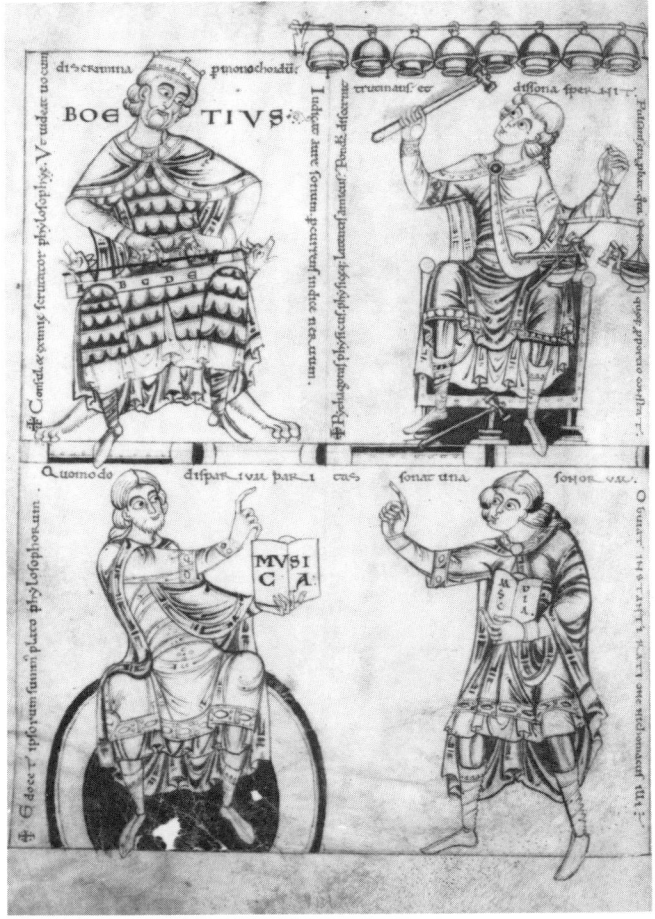

Figure 11 Great mathematicians of the distant past: Boethius, Pythagoras, Plato and Nicomachus, as portrayed in a mediaeval manuscript from the University of Cambridge.

By the time of Boethius, the Western Roman Empire had collapsed. Some four years after Boethius' death, in AD 529, the Eastern Emperor Justinian ordered the closure of the Academy, as the Neoplatonism it taught was seen as antagonistic to Christianity. The teachers moved east to the court of the Persian King Chosroes. Although they came back to Athens after a few years, the living succession of Neoplatonic learning had been impaired. One of these teachers was the commentator Simplicius, to whom we owe the preservation of the fragment of Eudemus about Hippocrates' work on lunes which you studied in Section 2.2.

So we can see the sixth century as the one in which the essentially Greek tradition of learning and commentating moved into yet a further phase. As a coda, we sketch briefly how the texts and knowledge of this tradition survived, in part, until the revival of classical learning in Renaissance Europe nearly a millennium later.

In simple terms we can think of there being three streams. The first is the continuing tradition within western Europe, initially in monasteries and then in universities. While the Dark Ages were by no means as dark as they have sometimes been represented, it cannot be claimed that a tradition headed by the works of Boethius transmitted much of mathematical consequence, beyond a simplified version of the *Arithmetic* and (which is much the same thing) the music theory of Nicomachus. Secondly, we have the very much richer and more complicated creative transmission via the world of Islam. This spread with astonishing speed and energy, after the death of Mohammed in 632, so that within thirty years all of the Greek-speaking communities along the southern and eastern shores of the Mediterranean were incorporated, as well as Mesopotamia and Persia. The subsequent Islamic assimilation of peoples and cultures included the translation into Arabic of vast numbers of Greek mathematical, scientific and philosophical texts. This Arabic heritage will be discussed in more detail in *Unit 5*.

The third stream of transmission or preservation was in that part of the old Eastern Roman Empire which held off the expansion of Islam for 800 years: the Byzantine civilisation, centred on Constantinople (now Istanbul). Only here, by and large, were the original texts preserved through copying in the original language. Of course, what survived in this way was only what was thought worth copying, for pedagogical or other reasons. Many ancient texts and also perhaps whole research traditions or mathematical topics did not survive this winnowing process generation by generation. Compared with the Arabs, and also with the earlier Hellenistic commentator tradition, the Byzantine scholars seem on the whole to have copied faithfully rather than developed the mathematics itself or creatively rewritten the texts. This is an advantage from our point of view while giving the impression of a curiously inanimate style of scholarship. One historian has written, half-sympathetically:

> The dead hand of academicism was laid upon them from the start by their determination to write—and, one suspects, to think—in a dead language, the Attic dialect of the fifth century BC; the weight of their classical inheritance crushed them.

Richard Jenkyns, *The Sunday Times*, 19 June 1983. (The Attic dialect was that of Plato and other Athenians of his time.)

Perhaps it was the Byzantine scholars of whom W. B. Yeats was thinking when he wrote, in his poem *Byzantium* (1930):

> Before me floats an image, man or shade,
> Shade more than man, more image than a shade;
> For Hades' bobbin bound in mummy-cloth
> May unwind the winding path;
> A mouth that has no moisture and no breath
> Breathless mouths may summon;
> I hail the superhuman;
> I call it death-in-life and life-in-death.

However, we should not overlook some Byzantine developments beneficial to later scholarship. One such was the invention, around 800, of spaces between words, and upper- and lower-case letters. Previously, Greek texts had no such differentiation or spacing, and presented an appearance here explained in the words of the historian of Greek mathematics David Fowler.

SURVIVINGEARLYGREEKPAPYRIARESETOUTJUSTLIKETHISAN
DINSCRIPTIONSARESOMETIMESCARVEDBOUSTROPHEDONICAL
LYBACKWARDSANDFORWARDSONALTERNATELINESINDIVERSDIA
LECTSINALIVINGCHANGINGVARIETYOFSTYLESASSOCIATEDWI
THLEGALADMINISTRATIVELITERARYSOCIALANDTECHNICALSUB
JECTSANDINANEVOLVINGANDVARYINGSTYLEOFHANDWRITINGS

Almost all the surviving Greek texts from Byzantium are written in the new style; that is, they date from the ninth century or later.

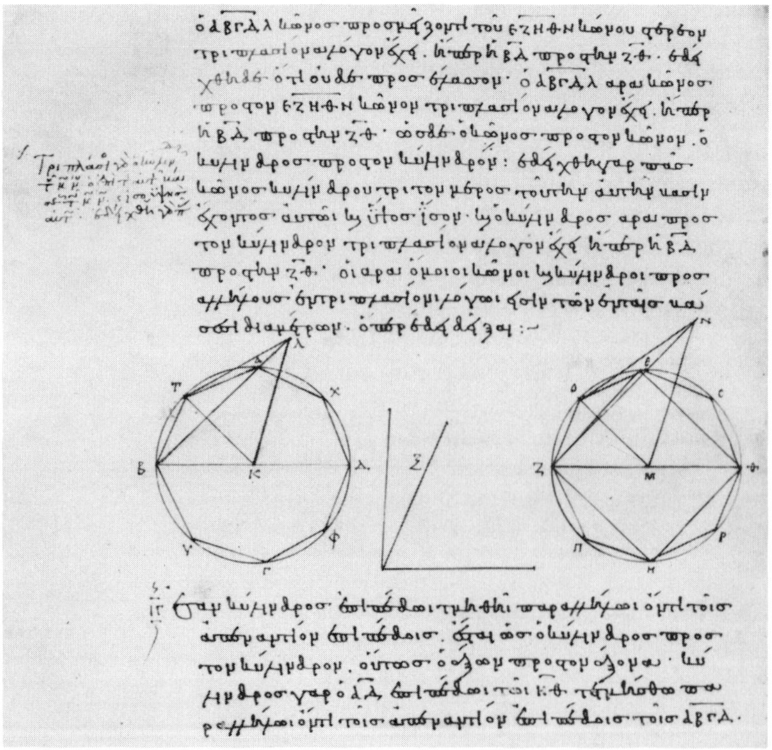

Figure 12 A page from the earliest surviving complete manuscript of Euclid's *Elements*, now in the Bodleian Library, Oxford. It was written, in the new Byzantine 'minuscule' script, by the scribe Stephen in AD 888.

To gain a more detailed understanding of the Byzantine transmission of texts, let us take as an example the works of Archimedes. Please *read the account in the Source Book by Sir Thomas Heath*, 'The Text of Archimedes' (**SB** 4.B4).

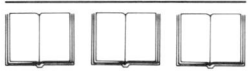

This shows several points of interest. Only in ninth-century Constantinople were 'the works' of Archimedes—those that Leon could find—collected together, and from that activity stems, basically, the text and knowledge of all subsequent generations. That manuscript found its way to the West—to Sicily—in the twelfth century, eventually disappearing some time in the middle of the sixteenth century. But notice too that a very important work by Archimedes, *The Method* (which we look at in *Unit 4*), was discovered only in 1906. It is possible that there are other lost texts still to be found, either in Istanbul or elsewhere. From the various dates given by Heath you can see that the fifteenth and sixteenth centuries were particularly active in the translating, editing and printing of the works of Archimedes. This process of the gradual spreading of knowledge of the work of Archimedes is fairly typical in general outline: a manuscript or two filtering through to western Europe in the later Middle Ages, then the trickle turning into a flood in the half-century leading up to and after the eventual fall of Constantinople to Islam (1453). The excitement among scholars, and the additional knowledge stimulated by the wealth of texts in Byzantine custodianship, were critical factors in what we call the Renaissance.

Here our initial survey of Greek mathematics, its history and transmission, ends. You have seen that the Byzantine civilisation preserved that culture (or the aspects of it we are interested in), up until the fifteenth century. Later we pick up the story of what happened in the Renaissance and onwards, but for the next two units we go back to the first two centuries of Greek mathematics—that is, from the decades before Plato (the latter part of the fifth century BC) to the time of Apollonius (later third century BC). This is because it was the mathematics from this period that the later Renaissance mathematicians devoted their energies to learning, digesting and generally coming to grips with, and which provided the springboard for the great flourishing of European mathematics in the seventeenth century. (Two later Greek mathematicians, Diophantus and Pappus, were also significant influences on particular seventeenth-century mathematicians, as we shall see in due course.) So we have a two-fold interest in classical Greek mathematics: it is of great interest in its own right, as a different style of mathematics from anything we know of previously, and it was the study of, reaction to, and development of these works that helped catalyse and stimulate the development of 'modern' mathematics three or four hundred years ago.